HIV and AIDS

David R. Wessner
Davidson College

Michael A. Palladino, Series Editor
Monmouth University

San Francisco   Boston   New York
Cape Town   Hong Kong   London   Madrid   Mexico City
Montreal   Munich   Paris   Singapore   Sydney   Tokyo   Toronto

Acquisitions Editor: Susan Winslow
Sr. Marketing Manager: Scott Dustan
Production Supervisor: Shannon Tozier
Production Management: Schawk, Inc.; Publishing Solutions for Retail, Book, and Catalog
Composition and Illustration: Schawk, Inc.; Publishing Solutions for Retail, Book, and Catalog
Manufacturing Buyer: Michael Early
Text Designer: Schawk, Inc.; Publishing Solutions for Retail, Book, and Catalog
Cover Designer: Seventeenth Street Studios
Cover Credit: Photo Researchers, Inc.
Cover Image: Scanning electron micrograph (SEM) of a whole T-lymphocyte infected with Human Immunodeficiency Virus (HIV). Small spherical virus particles visible on the surface are in the process of budding away from the cell membrane.

ISBN 0-8053-3956-6

Copyright ©2006 Pearson Education, Inc., publishing as Benjamin Cummings, 1301 Sansome St., San Francisco, CA 94111. All rights reserved. Manufactured in the United States of America. This publication is protected by Copyright and permission should be obtained from the publisher prior to any prohibited reproduction, storage in a retrieval system, or transmission in any form or by any means, electronic, mechanical, photocopying, recording, or likewise. To obtain permission(s) to use material from this work, please submit a written request to Pearson Education, Inc., Permissions Department, 1900 E. Lake Ave., Glenview, IL 60025. For information regarding permissions, call (847) 486-2635.

Many of the designations used by manufacturers and sellers to distinguish their products are claimed as trademarks. Where those designations appear in this book, and the publisher was aware of a trademark claim, the designations have been printed in initial caps or all caps.

9 10 — V036— 10
www.aw-bc.com

Contents

An Introduction to HIV/AIDS — 1

HIV/AIDS Basics — 2
- *What Is a Virus?* 2
- *What Is HIV?* 4
- *What Is the Immune System?* 9
- *What Is AIDS?* 10
- *How Is HIV Transmitted?* 18

Treatment — 21
- *Antiretroviral Drugs* 21
- *Antiretroviral Therapy* 24

Search for a Vaccine — 25
- *Vaccine Basics* 25
- *HIV-Vaccine Candidates* 27
- *Why Don't We Have a Vaccine?* 28

Current Issues — 29
- *Access to Antiretroviral Drugs* 29
- *Abstinence Versus Condoms* 30
- *Needle-Exchange Programs* 32
- *Global Fund to Fight AIDS, Tuberculosis, and Malaria* 33

What Does the Future Hold? — 33

Resources for Students and Educators — 36
- *For Students* 36
- *For Educators* 38
- *Journal Articles* 38
- *Popular Books* 38

An Introduction to HIV/AIDS

In the summer of 1981, a new cable channel devoted to music videos, MTV, went on the air; Ronald Reagan was beginning his first term as president of the United States; and the first reports of the disease now known as **AIDS,** or *a*cquired *i*mmuno*d*eficiency *s*yndrome, appeared in the scientific literature.

In June of 1981, Dr. Michael Gottlieb and colleagues published a short report in *Morbidity and Mortality Weekly Report* (MMWR) describing a group of male patients treated for pneumonia caused by the bacterium *Pneumocystis carinii*. To add to the mystery, the authors noted that all of the patients were gay men and exhibited signs of a severe immunodeficiency, leading to the speculation that a new, sexually transmitted pathogen could be responsible for this disease. And, as the editors of MMWR noted, *P. carinii* infections in young, previously healthy individuals were unusual. Most likely, few people, if any, imagined that this short report in MMWR was the first documentation of a major pandemic that would affect the world in a dreadfully tragic way.

Since 1981, researchers have learned a great deal about AIDS as a disease and about the *h*uman *i*mmunodeficiency *v*irus (**HIV**), the cause of AIDS. It is safe to say that we know more about HIV than any other virus. As a result of the unprecedented examination of this virus, over 20 drugs that are effective against HIV have been approved for use in the United States. Many people now believe that AIDS can be viewed as a chronic, manageable disease.

Yet, we still are faced with a global health crisis. An estimated 42 million people currently are infected with HIV. An additional 15,000 people become infected every day. Seventy percent of the people with HIV/AIDS live in sub-Saharan Africa. For most of these people infected with HIV, antiviral drugs are not available; indeed, for many of these people, adequate health care also is not available. And the ultimate preventative agent, a safe, effective vaccine, remains elusive.

In this booklet, we will examine various aspects of this horrific pandemic, including the biology of HIV, the immune system, and AIDS. We also will discuss many of the drugs currently available to combat AIDS, as well as the ongoing search for an effective vaccine. Finally, we will address important global issues associated with this pandemic, including access to affordable health care, distribution of condoms, and the empowerment of women. Throughout the booklet, key terms will appear in bold. In addition, useful listings of Web sites, journal articles, popular books, and references appear at the end of the booklet. Hopefully, this information will provide you with a greater understanding of HIV/AIDS and a greater appreciation of the need for all people to work together to end this pandemic.

HIV/AIDS Basics

WHAT IS A VIRUS?

When most of us hear the word *virus,* we may think about influenza, smallpox, or HIV. Indeed, these viruses have caused an immeasurable amount of human suffering and death. But, in fact, these human disease–causing agents, called **pathogens,** account for only a very small percentage of all the viruses found in nature. Some people estimate that there are over 10 million virus particles in every milliliter of sea water! Certainly, all of these viruses do not cause human diseases. In fact, most of these viruses do not even infect humans. Viruses that infect insects, plants, and even bacteria have been isolated and studied. There probably are viruses that infect every type of **eukaryotic organism** and **prokaryotic organism.** Before we examine the specific attributes of HIV, let's examine a more basic question: What is a virus?

Most notably, all viruses are small. Viruses typically are between 10 and 100 nanometers (1 nanometer = 10^{-9} meters) in diameter. Contrast the size of viruses with the size of typical prokaryotic cells (1 to 10 microns, where 1 micron = 1,000 nanometers) and typical eukaryotic cells (10 to 100 microns) to get a sense of how small viruses really are. Of course, all viruses possess a **genome,** which is the virus's genetic material. Unlike prokaryotic and eukaryotic organisms, the genomic material of viruses may be deoxyribonucleic acid (**DNA**) or ribonucleic acid (**RNA**). Additionally, some viruses possess **double-stranded genomes,** while other viruses possess **single-stranded genomes.** The genome size of viruses also are small, which is not a surprise. These genomes range in size from a few thousand base pairs for the smallest viruses to about 200,000 base pairs for the largest viruses. Contrast this with the genome size of the bacterium *E. coli* (4.6 million base pairs) and with the genome size of humans (3 billion base pairs). Clearly, viral genomes are quite economical!

All viruses also are **obligate intracellular parasites.** In other words, viruses replicate inside other cells and require the resources of the host cell. As shown in Figure 1, a major hurdle for all viruses is to get inside a host cell. To enter an appropriate host cell, most viruses bind to the cell through a specific interaction between a protein that is present on the surface of the virus (often called the **viral attachment protein**) and a protein that is present on the surface of the cell (often called the **receptor**). After this attachment event, the virus penetrates the cell membrane and enters the host cell. Within the cell, the viral genome is replicated and viral proteins are produced. Usually, the host cell's **transcription** and **translation** machinery are used to carry out these activities. These newly formed components then assemble to form new viruses. To exit the cell, some viruses bud from the host cell and acquire a lipid bilayer in the process. These viruses are referred to as **enveloped viruses.** Other types of viruses exit the cell by lysing, which is a process of cell rupturing; they do not acquire a lipid bilayer. These viruses are referred to as **nonenveloped viruses,** or naked viruses. In this manner, a single infected cell may produce up to 10,000 new viral particles.

After exiting the cell, new viral particles may bind to other host cells within the organism and repeat the process, or the newly formed viruses may be shed from the organism and transmitted to other susceptible individuals. **Respiratory viruses,** such as influenza, are shed through coughing and sneezing. If these airborne viruses are inhaled by another person, they may infect cells within the respiratory tract and begin a new infection. **Enteric viruses,** such as rotavirus, are shed through fecal material. These viruses may enter the water supply and be ingested by another person. The viruses then infect cells of the gastrointestinal (GI) tract and begin a new infection.

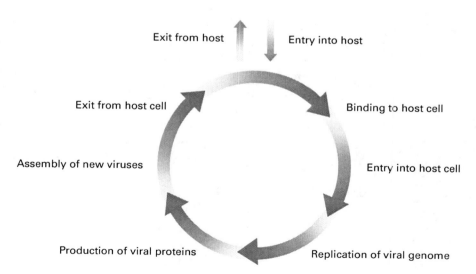

FIGURE 1. Schematic of the viral replication cycle.

HIV, as we will see, is a **bloodborne virus** that can be transmitted through bodily fluids, such as blood and semen. Thus, the site of entry, the appropriate host cells, and the site of exit for different viruses differ tremendously. Not surprisingly then, the types of disease associated with different viruses also differ tremendously. For all of these types of viruses, disease occurs when the virus begins destroying infected cells in the body. Influenza will cause respiratory problems by destroying cells of the respiratory tract, rotavirus will cause GI problems by destroying cells of the GI tract, and HIV will cause an immunodeficiency by destroying cells of the immune system.

WHAT IS HIV?

Most of us can provide a simple answer if asked, "What is HIV?" HIV is the virus that causes AIDS. And most of us also know that HIV is an acronym for human immunodeficiency virus. As the name tells us, HIV is a virus that infects humans and causes an immunodeficiency or malfunctioning of the immune system. Let's now look at HIV in greater depth.

HIV Structure and Replication

HIV is a virus in the **retrovirus** family of viruses. Retroviruses all share a number of characteristics. All retroviruses have a single-stranded RNA genome and are enveloped (see Figure 2). All retroviruses possess three very similar genes: (a) the *pol* **gene,** which encodes important viral **enzymes;** (b) the *env* **gene,** which encodes the major **envelope glycoproteins** (the sugar-coated proteins found embedded in the envelope); and (c) the *gag* **gene,** which encodes a series of proteins found inside the envelope. But most notably, all retroviruses have a very unusual replication cycle. Like most viruses, retroviruses begin their replication cycle by binding to and entering a host cell. But here is where things get a bit unusual. As we will discuss in detail later, the viral genome eventually becomes part of the host cell's genome. Then, the host cell machinery actually produces new viral proteins and new copies of the viral genome. These pieces assemble into new viral particles that are capable of infecting other cells or being transmitted to other individuals.

Now, let's fill in some of the details by looking at the steps in the HIV replication cycle, as shown in Figure 3. For HIV, the initial binding event involves a specific interaction between **gp120,** which is the major viral envelope glycoprotein produced by the *env* gene, and **CD4,** which is a protein found on the surface of certain human immune system cells (Step 1). Additionally, HIV must interact with a second protein on the host cell, usually CCR5 or CXCR4. Thus, CD4 is referred to as the **receptor,** and CCR5 and

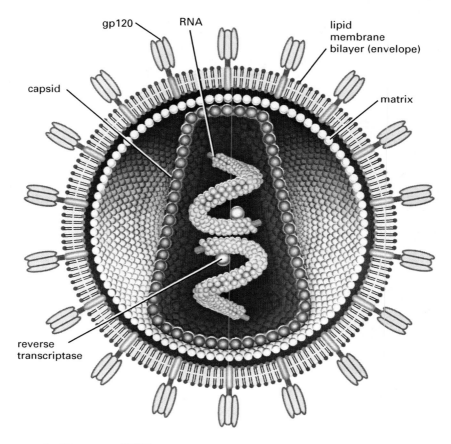

FIGURE 2. Structure of HIV.
Source: Adapted from the National Institute of Allergies and Infectious Diseases.

CXCR4 are referred to as **coreceptors.** After binding to and entering a host cell, the viral RNA is converted to DNA by the viral enzyme known as **reverse transcriptase** or RT (Step 2). This enzyme is so named because it seems to perform a function that is opposite of transcription; reverse transcription converts RNA to DNA. After the single-stranded viral RNA is converted into double-stranded DNA, it moves from the cytoplasm of the infected cell to the nucleus (Step 3.) Once the viral DNA is inside the nucleus, another unusual retroviral enzyme, **integrase,** cleaves (cuts) the host cell DNA and facilitates the insertion of the retroviral DNA into the host cell genome. At this point, the viral genetic material has become a permanent part of the infected cell's genome (Step 4).

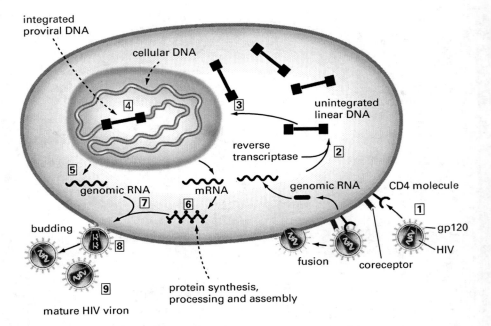

FIGURE 3. Schematic of the HIV replication cycle.
Source: Adapted from the National Institute of Allergies and Infectious Diseases.

Once the retroviral DNA has been integrated into the host cell's genome, the host cell's transcriptional machinery produces full length RNA copies of the viral DNA (Step 5). Viral genes are **transcribed** and **translated** by the host cell, resulting in the production of new viral proteins (Step 6). Within the cytoplasm, the newly produced viral RNA and proteins come together to form new viruses (Step 7), which **bud** from the host cell membrane, thereby exiting the cell (Step 8). One final step remains: Within these immature viral particles, the viral **protease** cuts the long, nonfunctional viral polypeptides produced by the host cell into smaller, functional viral proteins (Step 9). At this point, the newly formed viral particle is capable of infecting another cell. A fantastic animation of this process, which includes the specific target sites of antiretroviral drugs on the viral replication cycle, is available at the Hoffmann-LaRoche Web site (http://www.roche-hiv.com).

Discovery of HIV

In 1983, 2 years after the first reports of AIDS appeared in the scientific literature, Luc Montagnier and colleagues at the Pasteur Institute in France isolated a previously unidentified retrovirus from people with AIDS. The same virus was isolated shortly afterwards by Robert Gallo and colleagues at the National Cancer Institute in the United States. This virus, initially referred to as **LAV** by the French group and as **HTLV-III** by the American group, eventually became known as the human immunodeficiency virus (HIV), which, as previously noted, is the causative agent of AIDS.

But how do we know HIV causes AIDS? To answer this question, we need to review a little microbiology history. Today, we understand that microorganisms can cause diseases. In the 1800s, however, this idea was not readily accepted. In 1882, microbiologist Robert Koch developed a set of postulates that needed to be fulfilled to conclusively link a microorganism with a disease. **Koch's postulates** can be summarized as follows:

- The causative agent must be strongly associated with the disease.

- The causative agent must be isolated and propagated in the laboratory.

- The causative agent, when administered to another individual, must cause the disease.

All of these criteria have been demonstrated with HIV. First, HIV can be found in virtually every person with AIDS. Second, the virus routinely is isolated and propagated in virology laboratories. Third, there have been several documented cases of AIDS occurring in individuals who accidentally were exposed to purified HIV through a needlestick in the laboratory. Moreover, AIDS-like diseases occur in several different animal species when they are experimentally inoculated with HIV or an HIV-like virus. For an excellent overview of the evidence linking HIV and AIDS, read the article from the National Institute of Allergies and Infectious Diseases, "The Evidence That HIV Causes AIDS," which can be found at their Web site (http://www.niaid.nih.gov/factsheets/evidhiv.htm).

Origins of HIV

Another intriguing and important question to address about HIV regards its origins. The search for the origin of HIV has focused on genetic studies of HIV and related viruses called **simian immunodeficiency viruses (SIVs)** that infect various primates. Studies have revealed that SIVs are genetically very similar to HIV. Virtually all researchers now agree that HIV originated in Africa when an SIV strain was transmitted from its natural, nonhuman primate host to a human being. This transmission could have occurred in any number of ways. For instance, in parts of Africa, primates are hunted as a source of "bush meat." It is easy to imagine that, while hunting or butchering an animal, a person could have been exposed to blood from the animal.

Because SIVs, like HIV, are bloodborne viruses, the person, then, could have become infected with the virus. As we discussed previously, viruses generally are very species specific; that is, they only can replicate in a single species. However, in some cases, a virus may be able to replicate in a nonnatural host. Possibly such a virus could mutate to become well adapted to this new host organism. Such a transfer of a virus from an animal host to a nonnatural host, namely humans, is referred to as a **zoonosis** or a **zoonotic transfer.**

By comparing the sequences of HIV and various SIVs, researchers now believe that **HIV-1,** the primary HIV type present in the human population, originated when SIV_{cpz}, a simian immunodeficiency virus present in the chimpanzee species *Pan troglodytes troglodytes*, "jumped" from the chimpanzee to humans. Using sophisticated computer-modeling programs, researchers at the Los Alamos National Laboratory estimate that this transfer may have occurred around 1930. Other researchers, though, believe the transfer may have occurred much earlier.

Since the initial transfer from chimps to humans, HIV-1 has mutated to become better adapted to replicating in humans. It also has mutated into several genetically distinct strains. Today, three main groups of HIV-1 have been identified: groups M, N, and O. Group M HIV-1 is **pandemic** (an epidemic that exists worldwide). Groups N and O HIV-1 are much less prevalent. Researchers still are not sure if the three groups have resulted from a single transfer of SIV_{cpz} to humans or three independent transfers. Within the group M viruses, a number of genetically distinct subtypes, or **clades,** exist. As we will discuss later, this high degree of sequence variability represents a major hurdle in the development of an effective vaccine against HIV.

While HIV-1 is the predominant cause of AIDS, a second human immunodeficiency virus, called **HIV-2,** also exists. This virus is less virulent than HIV-1, causing a milder immunodeficiency. Sequence analyses indicate that this virus originated from an SIV naturally found in primates called sooty mangabeys. Currently, HIV-2 exists primarily in West Africa, although people with HIV-2 infections have been identified in many countries worldwide.

As we have seen, HIV originated from zoonotic transfers of SIV. It is quite likely that the transfer of various SIVs to humans have occurred repeatedly throughout human history. In most cases, though, these were dead-end transfers; the virus never became capable of replicating efficiently in humans or of being transmitted from one human to another. However, in the case of HIV-1, the transfer was highly successful. As the human population continues to grow and human communities continue to expand, contact between humans and wild animals certainly will increase. One wonders, then, if the emergence of HIV is an aberration or merely a harbinger of pandemics yet to occur.

WHAT IS THE IMMUNE SYSTEM?

Before we can understand fully how HIV causes an immunodeficiency, we need to understand the normal functioning of the human immune system. As we mentioned earlier, most viruses do not infect humans, and the viruses that can infect humans often do not cause disease. Even infections by viruses such as influenza and smallpox do not always result in disease. Why?

Most viruses have evolved to be very species specific and cell specific. In other words, they can replicate only within certain cells of a particular species. While a teaspoonful of sea water may contain millions of viral particles, many of these viruses may be capable of infecting only certain kinds of bacteria. Certainly, then, they will not cause us any problems! As we mentioned earlier, viruses typically only can replicate within certain types of cells. Rotavirus, for instance, only can infect intestinal cells. If this virus is inhaled, it will not encounter the appropriate host cells and, therefore, it will not be able to infect a cell.

But we are protected by more than just chance events. The human body possesses an amazing array of protective mechanisms against infectious agents. **Nonspecific resistance** protects us against all types of pathogens in a very general way. For example, physical barriers, such as our skin, the acidity of our stomach, and tear drops, protect us; these physical barriers can prohibit the entry of bacteria and viruses, they can disrupt the structure of bacteria and viruses, or they can wash bacteria and viruses away from vulnerable locations. The human **immune system** also possesses several nonspecific components: **Macrophages, neutrophils,** and **natural killer (NK) cells** are cells present in the **circulatory system** and the **lymphatic system.** These cells all are capable of recognizing and destroying a wide range of foreign invaders, including bacteria and viruses. These nonspecific resistance cells probably also identify and destroy tumor cells, thereby providing us with some protection against cancer.

The immune system also includes **specific resistance.** Unlike the nonspecific resistance, this portion of the immune system involves components that target specific pathogens. Specific resistance primarily involves two types of **white blood cells** that are also known as **lymphocytes: B lymphocytes (B cells)** and **T lymphocytes (T cells).** These T cells can be divided further into **T helper (T_h) cells** and **T killer (T_k) cells,** or **cytotoxic T lymphocytes (CTL) cells.** The T_h cells are characterized by the presence of the **CD4 protein** on their cell surface. These cells, therefore, often are called **CD4$^+$ cells** or **T4 cells.** The T_k cells, conversely, are characterized by the presence of the **CD8 protein** on their cell surface and are called **CD8$^+$ cells** or **T8 cells.**

B cells produce **antibodies** that provide **antibody-mediated immunity,** also known as **humoral immunity.** Antibodies are proteins secreted by the B cells into the blood that bind very tightly to foreign proteins, referred to as

antigens. The antibody-antigen interaction is very specific; a certain antibody only will interact with a certain antigen. Once a pathogen such as a virus, for example, becomes covered with antibodies, the pathogen is targeted for destruction by other components of the immune system. Antibodies also may neutralize a virus by covering the viral attachment protein, thereby preventing it from interacting with its receptor. The T_k cells provide **cell-mediated immunity.** Proteins on the surface of the T_k cells interact with foreign antigens present on the surface of a host cell, such as viral proteins present on the surface of an infected cell. As with the antibody-antigen interaction, this interaction is very specific. After the T_k cell has bound to a target cell, it releases a series of toxic chemicals that destroy the target cell. The T_h cells, as their name implies, help the other cells of the immune system. The T_h cells secrete **lymphokines,** factors that are necessary for the maturation and correct functioning of other components of the immune system, including B cells and T_k cells.

Another aspect of the immune system deserves mentioning—*memory.* Every time B cells and T cells interact with a specific antigen, a subset of the cells become **memory cells.** These cells are very long-lived, and they retain their "memory" of the specific antigen with which they initially interacted. If the same foreign antigen appears in the body again, after the body is reexposed to the same virus later in life, these memory cells rapidly proliferate and quickly begin to target the antigens, or foreign proteins. Often, this rapid response will prevent the pathogen from initiating an infection. As a result, people generally get certain diseases, like measles or chicken pox, only once in their lifetime. And, as we will soon discover, these memory cells also are responsible for the protective effects of vaccines.

The complex, amazing human immune system offers a great deal of protection from infectious agents. However, the system is not perfect. Despite the efforts of B cells and T cells, infectious agents occassionally overwhelm our defenses, resulting in disease.

WHAT IS AIDS?

From our discussions of HIV and the immune system, it may begin to be clear to you how HIV infection leads to an acquired immunodeficiency syndrome (AIDS). HIV infection destroys one of the key components of the immune system by delivering a knockout blow to T cells. Let's look at the steps in the process:

- HIV binds to and destroys $CD4^+$ cells.
- T_h cells are $CD4^+$ cells.

- Without T_h cells, B cells and T_k cells cannot function.
- Without functional B cells and T_k cells, our immune system fails.
- Without a strong immune system, we become susceptible to infections.

However, a person does not have AIDS as soon as he or she becomes infected with HIV. As with any infectious agent, it takes time for the virus to destroy enough host cells to cause disease. For example, think about food poisoning: If you have suffered through this disease, you certainly know that you do not get sick immediately after eating contaminated food. It takes time for the pathogen to replicate before it causes you to be sick. With HIV, unlike most infectious agents, it typically takes a long time before HIV results in AIDS. In most people who do not receive any treatment, AIDS occurs 6 to 8 years after the person initially became infected. This long interval between infection and disease is very unusual.

To gain a better understanding of the progression of HIV infection to AIDS, let's examine how one detects an HIV infection and how one determines if the HIV infection has progressed to AIDS.

Detecting HIV Infections

The determination of an HIV infection (**HIV positive**) in a person can be reached from two different approaches. The first approach is to look for the virus itself. The second approach is to look for the specific signs of a viral infection. Although the second approach is indirect, it is easier and was developed first. Therefore, let's explore the second approach first. As we discussed previously, our immune system reacts to every infectious agent that invades our body. Antibodies are produced and appear in the **serum**. Serum is the liquid portion of blood without the various cells and cell fragments (red blood cells, white blood cells, and platelets) normally present in blood. A person who has antibodies to a particular virus in his or her serum is said to be **seropositive** or to have **seroconverted.**

How can we use this immune response to determine if a person is HIV positive? If a person has been infected with HIV, the antibodies that react with HIV proteins will be present in that person's blood. One way to determine if a person has been infected, then, is to look for the presence of **anti-HIV antibodies** in that person's serum. The typical antibody test is an ***e**nzyme-**l**inked **i**mmuno**s**orbent **a**ssay* (**ELISA**). This assay involves a few simple steps, as outlined in Figure 4.

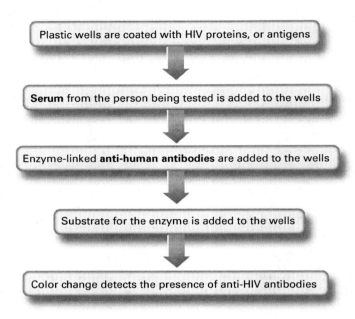

FIGURE 4. Detecting anti-HIV antibodies by ELISA.

If anti-HIV antibodies are present in the serum, they will attach to the HIV antigens placed in the wells and the enzyme-linked anti-human antibodies will bind to the human anti-HIV antibodies, thus creating a sort of antibody sandwich. The enzyme attached to the anti-human antibodies will convert its substrate into a colored compound. Thus, when the substrate is added to the wells, a color change indicates that all of the components are present in the well (see Figure 5). In other words, HIV antibodies were present in that person's serum. However, if there is no color change in the assay, this indicates that antibodies were not present. For a great tutorial and animations on detecting anti-HIV antibodies by ELISA, visit the Biology Project Web site of the University of Arizona (http://www.biology.arizona.edu/immunology/activities/elisa/main.html).

The HIV ELISA was developed in 1985, 4 years after the first scientific reports of AIDS and 2 years after the isolation and initial characterization of HIV. In March of 1985, officials began testing donated blood for HIV antibodies using this test. Soon thereafter, the U.S. blood supply was considered safe. Unfortunately, many people became infected with HIV from contaminated blood transfusions or contaminated blood products before this test was perfected. Today, the ELISA remains the major test for screening donated blood and it also is the initial test used to determine if an individual is HIV positive.

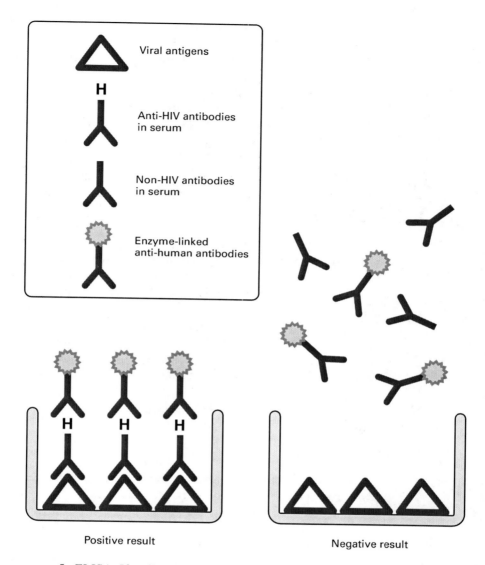

FIGURE 5. ELISA: If antibodies specific for the viral antigen are present in the serum, they will bind to the antigens, and the enzyme-linked anti-human antibodies will bind to these primary antibodies. When substrate is added, a color change will occur, indicating the presence of antiviral antibodies.

The ELISA is a very useful test for several reasons. Most notably, the test is easy to conduct and it is inexpensive. Unfortunately, the ELISA can produce **false positives,** where serum samples appear to be positive when they actually are negative. When a person receives a positive ELISA, the ELISA is repeated. If the second ELISA is negative, the person is assumed to be HIV negative. However, if the second ELISA also is positive, then a confirmatory **western blot** test is performed. This test also was designed to detect the presence of antibodies in the serum. In the western blot test, purified HIV proteins first are subjected to *polyacrylamide gel electrophoresis* **(PAGE).** In this procedure, the viral proteins are separated based on their size. These separated proteins, then, are tranferred to a thin sheet of nitrocellulose paper, which acts as a solid support that the proteins attach to. As in the ELISA, serum from the patient is added to the membrane.

If specific anti-HIV antibodies are present in the serum, they will bind to the HIV proteins (viral antigens). These bound antibodies are detected in much the same manner as they are in the ELISA. The western blot is more expensive and more time consuming than the ELISA. However, the western blot test has the advantage of resulting in fewer false positives, as well as providing more specific information. Rather than simply displaying a positive or negative well, as the ELISA does, the western blot test allows researchers to determine the exact size of the antigens to which antibodies are binding. These features make the western blot a good confirmatory test. When a person receives a positive western blot result, the person is considered to be HIV positive.

Even negative ELISA and western blot results, however, do not guarantee that a person is not infected with HIV. In rare cases, individuals do not make anti-HIV antibodies, even though they are infected. A bigger concern, though, is that there is a lag period between the time a person becomes infected and the time when sufficient levels of antibodies are present. For most people, this lag time is between 6 and 18 weeks; however, for other people, the lag period may be much longer. Therefore, a negative result may occur because the testing was done too soon after the initial infection. What does this mean for a person who may have been exposed to HIV either through risky sexual behavior, a shared needle, or some other means? Most officials recommend being tested for HIV after the potential exposure and again in 6 months. In most cases, a sufficient antibody response should have occurred during this period.

Of course, both the ELISA and the western blot are indirect tests; they both detect immune responses to HIV, not HIV itself. These tests also do not determine how much HIV is present in the blood (the **viral load**). As we will see, determining the level of HIV in the blood is very important in devising appropriate treatment strategies for an HIV-positive person. Several tests have been developed to detect HIV directly. Two of the most useful tests involve the use of a **polymerase chain reaction** (**PCR**) to detect the amount of HIV DNA or RNA in the blood. In this molecular technique, a specific fragment of DNA (in this case, HIV-specific DNA) can be amplified in a short period of time. A very good tutorial on PCR is available at the Cold Spring Harbor Laboratory Web site at http://www.dnalc.org/shockwave/pcranwhole.html. Remember from our earlier discussion of HIV replication that the viral RNA is converted into double-stranded DNA and becomes integrated into the host cell's genome. PCR can be used to detect this **proviral DNA** in the following manner:

- T cells are isolated from an individual.
- DNA is isolated from these cells.
- PCR is used to amplify proviral DNA.
- Amplified DNA is detected and quantitated.

Current versions of this test are so sensitive that HIV can be detected even if only 1 in 150,000 cells contains proviral DNA.

A similar direct test detects HIV RNA in the blood. Rather than determining how many T cells contain proviral DNA, though, this test provides an accurate determination of how many viral particles are present in the blood of an HIV-positive person. Initially, RNA is isolated from a blood sample. A form of PCR, referred to as **reverse transcription PCR (RT-PCR)** is used to amplify viral RNA. Reverse transcriptase (the very same enzyme used by HIV during its replication) converts viral RNA into DNA. PCR then is used to amplify this DNA. Again, the amount of viral RNA present in the blood sample can be quantitated. Current versions of this test can detect as little as 50 copies of viral RNA per milliliter of blood! Clinicians now believe that determining the amount of viral RNA in the blood provides them with the best indication of the overall health of their patients.

One major drawback of the traditional HIV ELISA has been the time required to get the test results. Generally, test results are not available until 1 to 2 weeks after the test is administered. Recently, several rapid tests have been approved for use by the Food and Drug Administration (FDA). These tests function in much the same manner as the ELISA, but the test results are available in 20 to 60 minutes. The **OraQuick test,** which was approved for use in 2002, detects HIV antibodies in a drop of blood obtained from a finger stick. In March 2004, the FDA approved a second form of the OraQuick test that detects HIV antibodies present in a swab of the mouth. Again, results are available in 20 to 60 minutes. Both of these tests can be conducted only in a physician's office. The FDA also has approved one at-home test. This test, called the Home Access Express HIV-1 Test System, allows a person to provide a drop of blood in the privacy of their home and to mail it to a designated laboratory for analysis. The person then can call a toll-free number to receive the test results. Many HIV/AIDS clinicians and advocates have expressed concern that people using such at-home tests often will fail to get necessary counseling after receiving their test results. It should be noted that a number of at-home tests are available over the Internet; however, none of these tests have been approved for use by the FDA, and their reliability should not be trusted.

Progression to AIDS

All of the tests described in the preceding section can be used to determine if a person is infected with HIV or they can be used to determine the viral load of an HIV-positive person. But, as we stated before, being HIV positive and having AIDS are not the same. What does it mean to have AIDS? In the early years of the pandemic, AIDS was defined by the presence of one or more **opportunistic infections** in a person. Opportunistic infections are infections caused by microorganisms that frequently are present but do not normally cause disease in humans. The most common opportunistic microorganisms

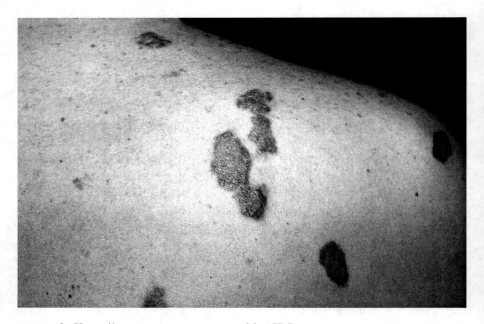

FIGURE 6. Kaposi's sarcoma on a person with AIDS.
Source: The University of Alabama at Birmingham, Departmentof Pathology PEIR Digital Library © (http://peir.net).

include the bacteria *Pneumocystis carinii,* the parasite *Toxoplasma gondii,* and the yeasts *Candida albicans* and *Mycobacterium avium.* Infections with these microorganisms cause a variety of diseases, including severe pneumonia (*P. carinii*) and oral thrush (*C. albicans*). Many people with AIDS also develop Kaposi's sarcoma, a cancer thought to be caused by a herpesvirus (see Figure 6). Interestingly, this disease seems to be much more prevalent in men who have sex with men than in other people with AIDS. Some people with AIDS also develop **AIDS-related dementia** or a severe **wasting syndrome,** or they may develop both conditions.

After the discovery of HIV, the definition of AIDS was modified to include evidence of both HIV infection and an opportunistic infection. Obviously, this definition still possessed some drawbacks. Most notably, the list of opportunistic infections could not be all-inclusive. Because the epidemic in the United States appeared first in gay men, the opportunistic infections generally associated with AIDS were diseases that appeared in this group of individuals. However, some researchers argued that women and infants, for example, could be prone to different opportunistic infections. In 1993, the **Centers for Disease Control and Prevention (CDC)** modified the definition of AIDS again to include HIV-positive people with a $CD4^+$ cell count of less than 200 cells per milliliter of blood. (Normal concentrations of $CD4^+$ cells are 800 to 1,000 cells per milliliter.) This CDC definition from 1993 remains in use today.

If we monitor the viral load, the antibody levels, and the $CD4^+$ cell count in an HIV-positive person, we can begin to understand the dynamics of this disease and how the progression to AIDS occurs. Before exposure to HIV, the $CD4^+$ cell count normally is 800 to 1,000 cells per milliliter of blood and HIV proteins and anti-HIV antibodies are not present in the blood. For the first few weeks after infection, the viral load is high and the anti-HIV antibody levels remain low. The $CD4^+$ cell count remains unchanged. During this period of **viremia** (the presence of high viral levels in the blood), the person may experience flu-like symptoms. Several months after infection, the viral load typically decreases dramatically, while the anti-HIV antibody levels rise significantly. Still, the $CD4^+$ cell count remains high, and the initial symptoms typically abate. This **asymptomatic period,** or period of **clinical latency,** may continue for several years. While it may appear that the infection has entered a latent phase, researchers now know that this phase is, in fact, very dynamic. Viral replication is occuring at a phenomenal rate, and the body is producing new $CD4^+$ cells at an equally phenomenal rate to replace the cells being destroyed by the virus. As currently estimated, over a billion HIV particles are produced each day during this asymptomatic period and over a million $CD4^+$ cells are destroyed and replaced each day!

At some point, the body begins to lose this battle. The CD4$^+$ cell count begins to decrease, and the viral load begins to increase. As the CD4$^+$ cell count decreases, the person's immune system becomes less effective. By the time the CD4$^+$ cell count has dropped to 200 cells per milliliter, the immune system is compromised, and the person becomes susceptible to numerous opportunistic infections. At this point, the person has progressed from being HIV positive to having AIDS. The compromised immune system and subsequent opportunistic infections, then, are the primary causes of most AIDS deaths.

HOW IS HIV TRANSMITTED?

The very first reports of AIDS published in scientific literature noted that the disease was present in gay men. This observation led **epidemiologists** to hypothesize that the causative agent was a **sexually transmitted** infectious agent. Subsequently, reports of AIDS in **injection drug users,** in people with hemophilia, and in blood-transfusion recipients caused epidemiologists to conclude that HIV also was **bloodborne.** Finally, reports of HIV in infants born to HIV-positive mothers suggested that **mother-to-child transmission** could occur. In this section, we will examine these three methods of transmission in more detail.

Sexual Transmission

We now know that HIV is present in all bodily fluids, especially blood, semen, vaginal secretions, and breast milk. Any activity that results in the transfer of these fluids, therefore, potentially can result in the transmission of HIV. We also know that HIV must replicate in CD4$^+$ cells, primarily T_h cells and macrophages, which are found in the blood. The question, then, is, what sexual activities permit the bodily fluids of an HIV-positive person to enter the bloodstream of another person?

Almost all sexual activities pose some risk. Unprotected penile-anal intercourse represents the most risky sexual activity. The rectum has a rich network of blood vessels, and the friction associated with penile-anal intercourse often causes slight, microscopic tears in the rectal lining. HIV present in the semen, then, can gain access to the circulatory system of the receptive partner. Likewise, unprotected penile-vaginal intercourse is a very risky activity. Again, the friction associated with penile-vaginal intercourse can cause tears in the vaginal lining, allowing HIV to enter the bloodstream of the female. Not surprisingly, the receptive partner in penile-anal intercouse or penile-vaginal intercourse is at greater risk. The inserting partner, though, also is at risk. HIV present in vaginal secretions or in the rectum can penetrate the mucous membranes of the penis, allowing transmission of HIV from an HIV-positive person to a male partner during penile-vaginal intercourse or penile-anal intercourse.

Other types of sexual activity are less risky, although certainly not risk free. Fellatio (oral sex performed on a male) represents a potential risk to both partners, although the performing partner certainly is at greater risk. If ejaculation occurs in the mouth, HIV present in the semen can enter the bloodstream through lesions in the mouth. In theory, HIV also can be transmitted by an HIV-positive person performing fellatio, especially if the person's mouth has cuts or sores. Similary, cunnilingus (oral sex performed on a female) represents a theoretical risk to both partners. Again, the risk involves the transfer of HIV present in vaginal secretions to tears in the mucosal membranes of the mouth. Most experts agree that the risk of HIV transmission from oral sex is fairly low and that cunnilingus is less risky than fellatio.

Very low-risk activities include mutual masturbation and deep kissing. Again, transmission of HIV is possible. If semen or vaginal secretions from an HIV-positive person come in contact with small abrasions in the skin of another person, there is a theoretical chance that the virus could be transmitted. Likewise, the virus could be transmitted through deep kissing, especially if both partners have lesions in their mouths. These risks, though, are very low.

Given our discussion of HIV transmission through sexual activities, prevention should be obvious: Prevent the transfer of bodily fluids! Condoms, when used correctly, greatly reduce the transmission of HIV. Similarly, the use of dental dams (latex barriers used during fellatio or cunnilingus) also are highly effective at blocking HIV transmission.

Bloodborne Transmission

Before an effective HIV test became available in 1985, many people became infected with HIV by receiving contaminated blood, either from blood **transfusions** or from blood products such as **factor VIII,** a blood-clotting factor used to treat hemophilia, which is a disorder characterized by the inability of the blood to clot properly. The HIV/AIDS epidemic in people with hemophilia has been especially shocking. Before the development and FDA approval of recombinant factor VIII in 1993, people with hemophilia were dependent on concentrated factor VIII obtained from large numbers of blood donors. By 1984, 90% of people with severe hemophilia and 50% of people with mild to moderate hemophilia had been infected with HIV through contaminated blood products!

Today, the U.S. blood supply is very safe. Through the use of extensive questionnaires for prospective donors and the testing of donated blood, the risk of acquiring HIV-infected blood has dropped tremendously, although some risk still does exist. Because of the possibility of false negative test results and the lag period between the time when a person becomes infected with HIV and when that person develops antibodies, some contaminated blood from donors not yet identified as HIV positive still exists within the available blood supply. The current risk, though, is very low.

While the rate of new HIV infections from blood transfusions and blood products has decreased dramatically, infections among injection drug users have increased dramatically. When needles are shared, blood from one user can be transferred to the next user. Not surprisingly, then, sharing needles represents a very effective means of transmitting HIV. Currently, about a third of all new HIV infections in the United States occur among injection drug users and their sexual partners. In several countries, such as Russia, the transmission of HIV by sharing needles probably accounts for the majority of new infections.

Mother-to-Child Transmission

Today, approximately 1,800 infants a day become infected with HIV through mother-to-child transmission. Most of these infections occur in sub-Saharan Africa. Generally, an HIV-positive mother transmits the infection to her infant during the birthing process or through breast feeding.

In developed countries, the rate of mother-to-child transmission has dropped tremendously for a few simple reasons. First, testing is readily available, so a pregnant woman can determine her HIV status. Second, as we will discuss in the next section, **antiretroviral drugs** also are readily available. Studies have shown that the administration of antiretroviral drugs can reduce dramatically the incidence of mother-to-child transmission. In the United States, clinicians currently recommend treating an HIV-positive pregnant woman with the antiretroviral drug **azidothymidine (AZT)** during the second and third trimesters of the pregnancy and during the delivery. In addition, clinicians also recommend treating the newborn infant of the HIV-positive woman immediately after birth. Studies have shown that even much shorter courses of treatment can reduce significantly the transmission of the virus. Third, for an HIV-positive woman, a Cesarean-section delivery is much less likely to result in the transmission of the virus than is a vaginal delivery. Finally, providing the infant with formula instead of virally contaminated breast milk eliminates another significant risk of mother-to-child transmission. Unfortunately, these options are not readily available in developing countries. Thus, in regions of sub-Saharan Africa, where as many as 40% of pregnant women may be HIV positive, the transmission of HIV to infants remains a terrible problem.

Treatment

ANTIRETROVIRAL DRUGS

In March of 1987, approximately 4 years after HIV was isolated, the first anti-HIV drug, which is also referred to as an antiretroviral drug, was approved for use in the United States by the Food and Drug Administration (FDA). As of August 2004, the FDA had approved 22 antiretroviral drugs, and more drugs currently are in development. A complete, up-to-date listing of the approved antiretroviral drugs is available on the National Institutes of Health Web site (http://aidsinfo.nih.gov/other/cbrochure/english/05_en.html).

It is worth noting that antiviral drugs are, in many respects, more difficult to design than are antibacterial drugs such as antibiotics. Because viruses rely on many of the normal cellular processes of their host cells for their own replication, it is difficult to design drugs that interfere specifically with viral replication, but do not significantly interfere with the replication of our cells. Bacteria, on the other hand, generally do not utilize the processes of our cells when they replicate. As a result, it is easier to develop drugs that interfere with bacterial cell replication, but do not affect the replication of our cells.

All of these antiretroviral drugs share one common property: They interfere with a specific aspect of the HIV replication cycle (see Figure 7). The first antiretroviral drug approved for use was azidothymidine (AZT). This drug is one of a class of available drugs that are known as **nucleoside analogs.** As the name indicates, these drugs are similar in structure to nucleosides, which are the building blocks of nucleic acids. Because these drugs differ slightly from nucleosides, though, they disrupt the process of DNA replication if they become incorporated into a growing strand of DNA. Our cellular DNA polymerase generally can differentiate these nucleoside analogs from the real nucleosides and avoid incorporating them into our DNA. The reverse transcriptase of HIV, though, is less discriminating; it routinely uses these nucleoside analogs when converting the viral RNA into DNA. As a result, viral replication is blocked.

Another class of antiretroviral drugs also interferes with the viral reverse transcriptase, but in a very different manner. **Nonnucleoside reverse transcriptase inhibitors,** or **NNRTIs,** bind directly to the viral reverse transcriptase and disrupt its activity. Again, the result is that viral replication is stopped. And because reverse transcriptase is an enzyme not normally present in our cells, these drugs do not disrupt important cellular functions.

In the mid-1990s, new antiretroviral drugs were approved by the FDA that do not interfere with reverse transcriptase. **Protease inhibitors,** or **PIs,** bind to and inhibit the activity of the viral protease. As we discussed earlier, the viral protease cuts the long, nonfunctional viral polypeptides present in newly formed viral particles into smaller, functional viral proteins. In

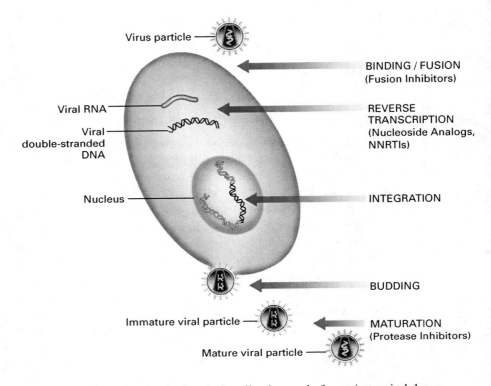

FIGURE 7. Sites of action in the viral replication cycle for antiretroviral drugs: Approved drugs and drugs currently in development target specific aspects of the viral replication cycle. The four major categories of antiretroviral drugs (nucleoside analogs, nonnucleoside reverse transcriptase inhibitors (NNRTIs), protease inhibitors, and fusion inhibitors) are indicated in brackets next to their site of action. Other drugs currently under development target integration and budding.

the absence of this enzyme, then, newly formed viral particles remain noninfectious.

The fourth class of antiretroviral drugs, **fusion inhibitors,** currently consists of only one drug. Enfuvirtide, also referred to as T-20 or Fuzeon, was approved for use in 2003. This drug interferes with the virus's ability to fuse with the cell membrane after the virus binds to its receptor. Obviously, if the virus cannot fuse with the cell membrane, then it cannot enter the cytoplasm, and replication cannot occur. Again, this drug interferes with a specific aspect of the viral replication cycle. Unlike drugs in the other classes, fusion inhibitors act at a much earlier stage of the replication cycle. Table 1 presents the four classes of antiretroviral drugs.

Currently, a number of new drugs are in development or are undergoing human trials. Some of these drugs fall into one of the four classes we have just described. Others target different aspects of the viral replication cycle. Most notable among these new drugs are compounds that target **integrase,** an essential viral enzyme, and the **budding** process that newly formed virus particles use to exit an infected cell. As with nucleoside analogs, nonnucleoside reverse transcriptase inhibitors, protease inhibitors, and fusion inhibitors, it is expected that these new drugs will inhibit viral replication effectively and that they also will have minimal effects on the normal workings of our cells.

One final note about these antiretroviral drugs: It is amazing how closely linked the development of therapeutic agents and basic science are. The development of these antiretroviral drugs was possible only because researchers had a good understanding of how retroviruses work. In fact, several researchers had been studying retroviruses long before the first cases of AIDS were reported, just because these unusual viruses represented an intriguing area of research. In the years before the AIDS virus was

TABLE 1. Classes of Available Antiretroviral Drugs

Generic name	**Brand or common name**	**FDA approval date**
Nucleoside analogs		
Zidovudine Azidothymidine	Retrovir, ZDV, AZT or Retrovir	March 1987
Didanosine	Videx, DDI	October 1991
Nonnucleoside reverse transcriptase inhibitors		
Nevirapine	Viramune, NVP	June 1996
Efavirenz	Sustiva, EFV	September 1998
Protease inhibitors		
Saquinavir	Invirase, SQV	December 1995
Ritonavir	Norvir, RTV	March 1996
Fusion inhibitors		
Enfuvirtide	Fuzeon, T-20	March 2003

discovered, some people might have argued that there was no need to understand how these obscure viruses replicated. Imagine how difficult antiretroviral drug development would have been without this knowledge!

ANTIRETROVIRAL THERAPY

Prior to the development of protease inhibitors, treatment for HIV involved the administration of one of the nucleoside analogs or one of the nonnucleoside reverse transcriptase inhibitors. This form of treatment is known as **monotherapy.** While treatment with AZT, for example, was effective at reducing the viral load and delaying the onset of AIDS, it clearly was not perfect. AZT has been a very effective antiretroviral drug; however, it has many side effects, including severe anemia and nausea. Furthermore, HIV mutates frequently, and HIV variants that are resistant to AZT have been isolated from many people. Because all of the nucleoside analogs function in basically the same manner, mutations that have resulted in AZT resistance often also have resulted in resistance to other drugs in this class.

Shortly after the approval of the first protease inhibitors, David Ho of the Aaron Diamond AIDS Research Center at the Rockefeller University recommended treating HIV-infected people with a nucleoside analog and a protease inhibitor simultaneously. The rationale, in retrospect, was simple: If we attack various aspects of the viral replication cycle simultaneously, then viral replication will be greatly inhibited; viral loads will decrease dramatically; and the development of resistant mutants will be much less likely. The results were remarkable. The viral loads in people infected with HIV dropped dramatically, T-cell counts improved, and the progression from HIV infection to AIDS was delayed significantly. For developing this improved method of HIV treatment and for his other contributions to HIV research, David Ho has received numerous awards and was named *Time* magazine's Person of the Year in 1996. To learn more about Ho and his research interests, visit the Aaron Diamond AIDS Research Center Web site (http://www.adarc.org/research/ho/index.htm).

Today, the recommended therapy for people with HIV involves the administration of three drugs, typically two nucleoside analogs and a protease inhibitor. This treatment, which is referred to as **highly active antiretroviral therapy (HAART),** is very effective. Recommendations of specific drug combinations can be accessed at the National Institutes of Health Web site (http://aidsinfo.nih.gov/guidelines/adult/AA_032304.html). For many people infected with HIV, HAART reduces the viral levels in the blood to undetectable levels and it also may delay the progression to AIDS indefinitely. Some clinicians now believe that HIV can be regarded as a manageable chronic disease.

While HAART has proven to be very effective in the treatment of HIV infections, it is not the perfect solution. Until recently, the required drug regimen was very complex. A person needed to take a large number of pills at different times of the day, some with food and some on an empty stomach. For many people, adhering to this regimen was difficult. Currently, a number of pharmaceutical companies are developing forms of the existing drugs that can be taken with greater ease. Several adverse side effects have been associated with antiretroviral therapy. Recently, clinicians have observed that people receiving HAART often develop metabolic problems, including elevated cholesterol levels and unusual fat accumulations. In addition, the development of drug-resistant variants of HIV remains a large problem. In a study published in 2002, for instance, researchers reported that nearly 25% of people newly infected with HIV were infected with a virus that was resistant to at least one of the commonly used antiretroviral drugs. Finally, and perhaps most important, the cost of the antiretroviral drugs make them inaccessible to most people throughout the world. In the United States, the costs of antiretroviral treatment easily can exceed $12,000 per year. We will discuss the worldwide implications of, and possible solutions to, this economic problem later in this booklet.

Search for a Vaccine

While the many antiretroviral drugs now available may allow clinicians to view HIV infection as a manageable chronic disease, these drugs are not the ultimate solution. Antiretroviral drugs do not "cure" a person's HIV infection; they merely decrease the virus's ability to replicate and cause disease. Clearly, a more desirable outcome would be the complete prevention of HIV infections. Safer sex practices could decrease dramatically the transmission of HIV, but this depends on the consistent participation of all sexually active people. An effective vaccine, on the other hand, potentially could provide long-term protection to all people. Unfortunately, this holy grail of HIV research still remains elusive. In this section, we will exmaine the biology of vaccines and the search for an effective HIV vaccine.

VACCINE BASICS

The human immune system is truly amazing. As we discussed previously, a two-pronged attack is unleashed against every infectious agent that enters the body. Furthermore, the immune system has a memory; after an initial exposure to an infectious agent, it is primed to attack the same agent swiftly and strongly upon a second exposure. In a way, then, we get vaccinated every time we are infected with a virus. The human body develops a primary response to the virus and, more important, prepares itself for a second infection. As a result, we get many viral diseases, such as chicken pox and measles, only once. After that initial exposure, the body develops an **immunological memory** and effectively combats the virus upon reexposure.

The theory of vaccines, then, is quite simple: By intentionally introducing a pathogen into the body, we cause the body to form a primary immune response and, more important, to develop an immunological memory of the pathogen. Edward Jenner used this approach over 200 years ago to vaccinate people against smallpox. Jenner noticed that milkmaids often acquired cowpox, a mild smallpox-like disease, but appeared to be resistant to smallpox. He hypothesized that the milkmaids were exposed to cowpox while milking cows. Exposure to the cowpox agent, he reasoned, conferred in the milkmaids protection against the structurally similar smallpox agent. To test his theory, he experimentally inoculated people with scrapings from the lesions on milkmaids. These vaccinated people developed resistance to smallpox.

Today, vaccines exist for a large number of viruses. All of these vaccines are based on the same simple model: Exposure to a structurally similar but nonpathogenic form of an infectious agent will result in the development of immunological memory that enables the body to effectively combat the pathogenic form of the infectious agent.

Generally, viral vaccines fall into one of three categories: attenuated, inactivated, or subunit. **Attenuated vaccines** involve the use of whole virus particles that are capable of replicating in humans, but these whole virus particles have been altered so that they are nonpathogenic (attenuated). These vaccines typically provide the strongest and most complete protection against infection. However, because these vaccines involve viruses capable of replicating, there is a risk associated with these vaccines. It is possible for the attenuated virus to mutate and become pathogenic. In addition, such vaccines are not appropriate for use in immunocompromised people. **Inactivated vaccines** involve the use of whole virus particles that have been rendered incapable of replicating. Usually, the inactivation is achieved by treating the virus particles with harsh chemicals or with high temperatures. Inactivated vaccines usually provide a weaker immune response than attenuated vaccines. Inactivated vaccines have the advantage, though, of increased safety; because the virus particles are not capable of replicating, there should be no threat of disease associated with the vaccine (unless, of course, the inactivation process was not complete). **Subunit vaccines** do not involve the use of whole virus particles. Instead, subunit vaccines are composed of one or more viral proteins. An immune response to these viral antigens develops, and it is this immune response that confers protection against the virus. An obvious downside to subunit vaccines is that the immune system does not "see" the entire pathogen. Instead, the immune response is limited to the antigens used in the vaccines. Subunit vaccines, though, are very safe. Because only viral proteins are used, there is no chance of viral replication occurring. Table 2 lists some representative viral vaccines that are approved for use in the United States.

HIV-VACCINE CANDIDATES

Because of the potential safety issues, many people are wary of attenuated or inactivated HIV vaccines. In fact, relatively little research into these types of vaccines has been conducted, although attenuated vaccines have shown some potential in monkeys. Instead, most HIV-vaccine research has focused on subunit vaccines. Usually, these subunit vaccines involve gp120 preparations. Because gp120 is the major HIV surface protein and the viral attachment protein, it seems reasonable that an immune response to this constituent of the virus would be most effective.

By the end of 2000, over 60 different candidate vaccines had been tested in humans. As of 2004, though, only one candidate vaccine progressed to **Phase III human trials.** In 1998, the FDA approved a Phase III trial of AIDSVAX, a gp120-based subunit vaccine candidate produced by VaxGen. These trials enrolled 5,000 people in the United States and 2,500 people in Thailand. Approximately two thirds of the people received the AIDSVAX preparation and the other one third received a placebo injection. These people all were HIV negative at the time they enrolled in the trial, and their HIV status was monitored for 3 years. At the end of the trial, it appeared that the

TABLE 2. Representative Viral Vaccines Approved for Use in the United States

Category of U.S. approved viral vaccines		
Attenuated	*Inactivated*	*Subunit*
Measles	Hepatitis A	Hepatitis B
Mumps	Poliovirus	
Chicken pox	Rabies	
Rubella (German measles)	Influenza (flu)	

AIDSVAX vaccine offered no protection against HIV infection. Equal percentages of the people who received the vaccine or the placebo became infected with HIV.

Today, many candidate vaccines involve very innovative strategies. Several researchers are exploring the use of recombinant virus vaccines. In these vaccines, a virus that does not cause disease in humans is genetically engineered to contain HIV genes. The rationale is that the virus can undergo limited replication in humans, like an attenuated vacccine, but can produce only one or two HIV proteins, like a subunit vaccine. Other researchers are exploring the effectiveness of DNA vaccines. In this approach, the DNA of selected HIV genes are injected into subjects in hopes of offering protection against HIV. In addition, the views of a successful vaccine are changing. Many people now believe that a vaccine does not need to provide 100% protection to be considered successful. Even if a vaccine offers only partial protection, or perhaps only inhibits HIV replication, thereby delaying the onset of AIDS, the vaccine still would be effective in the fight against HIV/AIDS. A good overview of HIV-vaccine research is available at National Institute of Allergy and Infectious Diseases (http://www.niaid.nih.gov/daids/vaccine).

WHY DON'T WE HAVE A VACCINE?

In 1984, when Robert Gallo and colleagues at the National Cancer Institute announced their discovery of HIV, Margaret Heckler, then secretary of the Department of Health and Human Services, boldly stated that a vaccine against the virus should be available by 1986. Now, in 2005, a vaccine still is not available, and most researchers doubt that an effective vaccine will be available in the near future. Why?

The answer is not simple. Certainly, much research into the development of an HIV vaccine has been conducted, and many candidate vaccines have been or are being tested. None of these candidate vaccines, though, appear particularly promising. One major obstacle to the development of an effective HIV vaccine is that HIV destroys the very cells necessary for a strong immune response. Another obstacle is the absence of a good animal model for vaccine studies. Probably the greatest obstacle to the development of an effective HIV vaccine is the great variability exhibited by HIV. HIV mutates rapidly and extensively. As a result, many genetically distinct forms of the virus, which as mentioned previously are referred to as clades, exist throughout the world. The gp120 proteins present in viruses of different clades vary enough from one another that the antibodies that react strongly with one gp120 protein may not react strongly with the gp120 proteins from another clade. Despite these difficulties, researchers remain hopeful that an effective HIV vaccine will be developed.

Current Issues

ACCESS TO ANTIRETROVIRAL DRUGS

As we discussed previously, the United States has approved 22 antiretroviral drugs for use against HIV infection. Through the correct use of these drugs, people with HIV can delay the progression to AIDS and lead relatively healthy lives for a number of years. As a result of these effective drugs, many clinicians now believe HIV can be considered a manageable chronic disease. However, as we also mentioned previously, these drugs are very expensive. In the United States, many people, although certainly not all people, can rely on private health insurance or government assistance programs to offset the costs of these drugs.

In much of the world, though, the costs of these drugs are beyond the reach of all people. At the XV International AIDS Conference in 2004, the official meeting slogan was "Access for All." Conference participants and

FIGURE 8. Protesters at the XV International AIDS Conference in 2004: Many participants at this conference argued that antiretroviral drugs should be made available to all people infected with HIV.
Source: Photo © David R. Wessner

protestors urged pharmaceutical companies to provide antiretroviral drugs to all people living with HIV, regardless of their ability to pay, as shown in Figure 8. In 2003, the World Health Organization and UNAIDS announced the **3 by 5 Initiative,** which is an ambitious plan to provide antiretroviral drugs to 3 million people living with HIV/AIDS in developing countries by the year 2005. Brazilian officials, in 2001, announced that they would allow companies in their country to produce **generic** versions of antiretroviral drugs, in violation of international patent laws. Several other countries, including India and Thailand, quickly followed suit. These generic drugs could be sold at a fraction of the cost of the brand name drugs. Pharmaceutical companies protested these actions. The production and sale of generic drugs, company officials argued, would prevent the companies from recouping the money they spent on research and development. Some officials also argued that there would be only limited quality control of these generic drugs. Incorrect formulations, they argued, could become available that might actually exasperate the HIV/AIDS problem. In 2003, the World Trade Organization reached an agreement with member countries that allows developing countries to produce or import generic versions of drugs still under patent protection. This agreement was reached to help combat public health epidemics, and the agreement certainly represents a major advance for the people of these countries.

ABSTINENCE VERSUS CONDOMS

As we stated earlier, condoms are highly effective in preventing the sexual transmission of HIV. In 1986, the Ugandan government recognized the severity of the HIV/AIDS epidemic within its country and began an extensive education campaign. Promoting condom use continues to be a major component of this campaign. Similarly, the Thai government began a major education campaign in the late 1980s designed to achieve 100% condom use among sex workers. In both countries, these education programs, which were based strongly on condom use, were largely successful. The incidence of new HIV infections in both countries decreased significantly.

Since the early 1980s, discussions about condoms have become much more commonplace. It no longer is unusual to see advertisements for condoms in magazines and on television. Condom usage is mentioned with increasing frequency on popular television shows. HIV/AIDS advocates often hand out condoms at bars and nightclubs to promote safer sex, and some individuals and groups are using highly creative means to get the safer sex message to as many people as possible, as shown in Figure 9.

FIGURE 9. Condom dress: Designed by Adriana Bertini, this dress, which was made entirely of condoms, was created to increase awareness of condoms and safer sex practices. Additional images and information are available at http://www.adrianabertini.com/br.
Source: © Reuters/CORBIS

Condoms also are highly controversial. In the United States, conservative groups have protested the distribution of condoms to young people or to students in schools. Some people argue that the availability of condoms promotes premarital sex. Abstinence, they say, is a better method of preventing HIV transmission. Currently, the U.S. government also favors abstinence over condoms, despite the clear evidence that increased condom use leads to decreased numbers of new HIV infections. In 2004, President George W. Bush drafted the **President's Emergency Plan for AIDS Relief (PEPFAR),** which is a 5-year plan for fighting HIV/AIDS worldwide. While PEPFAR represents a much needed commitment by the United States to combat the AIDS pandemic, many critics have argued that it focuses too heavily on abstinence. Certainly, abstinence does prevent the sexual transmission of HIV. Many people argue, though, that promoting abstinence is not the most effective means of preventing the spread of this virus.

NEEDLE-EXCHANGE PROGRAMS

While unprotected sex may account for a majority of new HIV infections, the sharing of needles among injection drug users also represents a major source of new infections. In fact, a report by the CDC in 2000 estimated that 25% of new infections in the United States occurred among injection drug users. If we then consider the sexual partners of injection drug users, as many as one third of all cases of HIV can be traced to the sharing of contaminated needles. In other countries, such as Russia and Ukraine, the sharing of contaminated needles represents the major source of new infections. In fact, a report by the Center for Strategic and International Studies estimated that, in 2000, 90% of HIV cases in Russia occurred in injection drug users and their sexual partners. Clearly then, strategies to decrease the transmission of HIV among injection drug users must be considered.

To combat the spread of HIV among injection drug users, a number of cities in the United States and other countries have initiated **needle-exchange programs.** In these programs, injection drug users can exchange used needles for clean, sterile needles. Advocates of these programs claim that needle-exchange programs lead to a decrease in the spread of HIV in a cost-effective manner. Opponents of these needle-exchange programs dispute their effectiveness in preventing HIV transmission and worry that the increased availabilty of needles could lead to increased drug use. Currently, the use of federal funds for such programs in the United States is banned.

GLOBAL FUND TO FIGHT AIDS, TUBERCULOSIS, AND MALARIA

Established in 2002, the **Global Fund to Fight AIDS, Tuberculosis, and Malaria** is a multinational partnership of public and private organizations aimed at preventing and treating these diseases. As stated on the Fund's Web site, these three preventable diseases primarily affect developing countries and cause "tremendous economic loss, social disintegration and political instability." Overseers of the Global Fund recognize that global cooperation is necessary in the fight against these diseases, with the financial resources of the developed countries being used to support programs designed by and for the developing countries.

In the first 2 years of its existence, the Global Fund pledged over $3 billion to support prevention and treatment programs in 128 countries. Despite this impressive funding record, activists have argued that developed countries, particularly the United States, have not contributed enough to the Global Fund. Indeed, many participants at the XV International AIDS Conference in 2004 argued that the United States should divert resources from PEPFAR to the Global Fund. While the United States has been a major contributor to the Global Fund, it also has earmarked a large portion of its HIV- and AIDS-related funds to PEPFAR. This arrangement, the U.S. government argues, provides the United States with greater flexibility in the use of these funds. Proponents of the Global Fund, on the other hand, argue that this arrangement hinders a cohesive global response to these diseases.

What Does the Future Hold?

As we have seen throughout this booklet, the HIV/AIDS pandemic is a public health problem of almost immeasurable magnitude that affects all nations of the world and all people of each nation. According to the World Health Organization, 40 million people worldwide were living with HIV at the end of 2003, 5 million people were newly infected in 2003, and 3 million people died of AIDS during that year. Over 20 million people worldwide have died of AIDS since the pandemic began in 1981, and an estimated 14 million children have been orphaned by AIDS. Today, nearly three quarters of the people living with HIV live in sub-Saharan Africa. In countries such as Botswana, Swaziland, and Zimbabwe, over one third of the people between the ages of 15 and 49 are HIV positive. And many experts fear that the epidemic is just beginning in countries such as China, India, and Russia. Obviously, to help, the response of the medical community and society in general needs to be of an equally immeasurable magnitude.

Luckily, our responses can be highly effective. In Brazil, the availability of antiretroviral drugs has resulted in a marked decrease in the yearly number of deaths due to AIDS. In Uganda, the government began a strong, comprehensive education program in 1986, and the HIV prevalence rates have dropped from 30% in 1990 to roughly 10% in 2000. And in Thailand, a government-sponsored education program with a goal of 100% condom use among sex workers has led to a dramatic decline in the HIV infection rate. Clearly, success is possible.

Prevention and treatment programs like the programs just noted need to reach all people at risk of being infected with HIV. Governments and non-governmental organizations must work together to ensure that programs are available for the groups that need them the most: the homeless, injection drug users, sex workers, and women and children. Many people in these groups, traditionally marginalized in many parts of the world, need to be empowered. Prevention and treatment programs must be designed in such a way as to be accessible to these groups and done in a way that eliminates the stigma, prejudice, and lack of basic human rights too often associated with the homeless, with injection drug users, and with women and children in many parts of the world.

These prevention and treatment programs are complemented by an increased scientific understanding of HIV/AIDS. Researchers continue to learn more about the virus and its effects on the body. This basic research is providing answers to the fundamental questions about HIV. Pharmaceutical companies continue to develop new antiretroviral drugs based on the advances in research. Drugs currently are being developed that target other steps of the HIV replication cycle, as are drugs that have less severe side effects. The search for an effective HIV vaccine continues. In the United States, the National Institutes of Health continue to provide nearly unprecedented funding for HIV/AIDS research. Worldwide, other national governments, the Global Fund, and private organizations like the Bill and Melinda Gates Foundation continue to fund important research. With continued funding and with the hard work of remarkable researchers, educators, and advocates throughout the world, important advances, and hopefully a cure, will result.

As the HIV/AIDS pandemic progresses through its third decade, the situation may look bleak; however, advances are occurring, and success stories do exist. Hopefully, this booklet has provided you with a good understanding of the biology of HIV/AIDS and the many issues associated with this disease. But a final question needs to be addressed: What can you do? First, and most important, protect yourself and your sexual partners. The spread of HIV is preventable, but we each need to do our part. Second, get involved. Talk about the prevention of HIV/AIDS with your friends. Organize educational outreach programs for your school or your neighborhood. Help raise funds for local, national, or international HIV/AIDS organizations. Participate in the annual National AIDS Marathon. Participate in a local AIDS walk. By working together, we can end the scourge of HIV/AIDS. As the slogan of the Nelson Mandela Foundation's 46664 campaign states, "Give 1 minute of your life to stop AIDS."

David R. Wessner, Ph.D.
Associate Professor of Biology
Davidson College
E-mail: dawessner@davidson.edu
Web site: http://www.bio.davidson.edu/people/dawessner

Resources for Students and Educators

FOR STUDENTS

46664 (http://46664.tiscali.com) The Nelson Mandela Foundation's Web site provides information about the foundation's campaign to fight HIV/AIDS.

ACT UP (http://www.actupny.org) ACT UP, the AIDS Coalition to Unleash Power, is one of the first, the best known, and the most influential HIV/AIDS activist groups.

AIDS Education Global Information System (http://www.aegis.org) This Web site has a worthwhile collection of many resources related to HIV/AIDS.

AIDSinfo (http://www.hivatis.org) The U.S. Department of Health and Human Services resource center provides a wealth of information about current and developing treatments for HIV/AIDS.

AIDS Marathon (http://www.aidsmarathon.org) If you are a runner, or want to become one, sign up for this annual fundraiser. Many cities have training programs designed for novices hoping to run their first marathon.

AIDS Walks (http://www.aidswalk.org) If running is not your thing, participate in a local AIDS walk. Organize a team or participate by yourself.

American Foundation for AIDS Research (http://www.amfar.org) This foundation is the nation's leading nonprofit organization for AIDS research, prevention, and treatment.

Avert (http://www.avert.org) This Web site offers a collection of varied materials, from current statistics to safer sex quizzes, about HIV/AIDS.

Bill and Melinda Gates Foundation (http://www.gatesfoundation.org) A Web site for one of the major contributors to HIV/AIDS research worldwide.

The Body (http://www.thebody.com) A great Web site for information about HIV/AIDS science, medicine, and policy. This site also includes a wonderful "Ask the Experts" section where clinicians and researchers answer questions posed by visitors to the site.

Centers for Disease Control and Prevention (http://www.cdc.gov/hiv) The CDC's Division of HIV/AIDS Prevention is at this Web site.

Gay Men's Health Crisis (http://www.gmhc.org) Officially founded in 1982, the GMHC has been one of the leading advocates of HIV/AIDS outreach and education.

The Global Fund to Fight AIDS, Tuberculosis, and Malaria (http://www.theglobalfund.org) This Web site provides information about the mission of the Global Fund and the programs that are currently funded.

History of HIV/AIDS Research at the NIH (http://aidshistory.nih.gov) This Web site has an interesting multimedia collection of materials that detail the early years of the pandemic.

HIV/AIDS in Popular Culture (http://www.bio.davidson.edu/projects/aidspopculture) This multimedia Web site displays portrayals of HIV/AIDS in art, music, film, and television.

HIV InSite (http://www.hivinsite.org) This very informative Web site is based at the University of California, San Francisco.

International AIDS Society (http://www.iasociety.org) The official Web site of the International AIDS Society, which is the organizer of the International AIDS Conferences.

Kaiser Family Foundation (http://www.kff.org) This Web site contains information on numerous health-related issues. A large section of this site is devoted to HIV/AIDS.

Know HIV (http://www.knowhivaids.org) A wonderful Web site aimed primarily at younger people, this site includes a list of upcoming television shows that deal with HIV/AIDS as well public service announcements about HIV/AIDS.

Living With AIDS (http://www.bio.davidson.edu/livingwithaids) This Web site has a short documentary that focuses on the support a woman living with HIV and her HIV-positive daughter receive.

National Institute of Allergies and Infectious Diseases (http://www.niaid.nih.gov/daids) The Web site of the NIAID's Division of AIDS includes up-to-date information about current research, funding programs, and clinical trials.

The New Mexico AIDS InfoNet (http://www.aidsinfonet.org) This informative Web resource was developed by the New Mexico AIDS Education and Training Center in the Infectious Diseases Division of the University of New Mexico School of Medicine.

New York Times (http://www.nytimes.com/aids) Posted on this Web site is a wonderful collection of HIV/AIDS articles that have appeared in the *New York Times* since 1981. This site also contains interesting audio and video files.

Office of AIDS Research (http://www.nih.gov/od/oar) The NIH's Web site provides information about budgetary, legislative, and policy elements of HIV/AIDS programs.

Red Hot (http://www.redhot.org) The Red Hot organization has been a leader in utilizing the recording industry to fight HIV/AIDS. The Web site has links to the many CDs and videos produced by this group.

Regional AIDS Interfaith Network (http://www.carolinarain.org) This is a diverse group of faith-based congregations who are devoted to providing care and support for people with HIV/AIDS.**UNAIDS** (http://www.unaids.org) The Web site of the Joint United Nations Programme on HIV/AIDS provides detailed information on UN programs focused on HIV/AIDS.

WHO HIV/AIDS Programme (http://www.who.int/hiv/en) The World Health Organization's Web site of HIV/AIDS resources.

World AIDS Day (http://www.worldaidsday.org) World AIDS Day is December 1 of each year. Go to their Web site to find out what's planned and how you can get involved.

XVth International AIDS Conference (http://www.aids2004.org) This Web site contains summaries of major developments from the 2004 conference in Bangkok, Thailand.

XVIth International AIDS Conference (http://www.aids2006.org) The XVIth International AIDS Conference Web site offers preliminary information about the 2006 conference to be held in Toronto, Canada in August 2006.

FOR EDUCATORS

CDC Division of HIV/AIDS Prevention (http://www.cdc.gov/hiv/pubs/mmwry.htm)
A collection of HIV/AIDS articles that have appeared in *Morbidity and Mortality Weekly Report*. This collection includes the very first reports of AIDS.

JOURNAL ARTICLES

Cohen, J. (2004). HIV/AIDS in Asia. *Science*, 304, 1931-1938.

Cohen, J. (2004). HIV/AIDS in China. *Science, 304,* 1430-1439.

Cohen, J. (2004). HIV/AIDS in India. *Science, 304,* 504-513.

Emini, E. A., & Koff, W. C. (2004). Developing an AIDS vaccine: Need, uncertainty, hope. *Science, 304,* 1913-1914.

Gallo, R. C., & Montagnier, L. (2003). Retrospective: The discovery of HIV as the cause of AIDS. *New England Journal of Medicine, 349,* 2283-2285.

Jaffe, H. (2004). Whatever happened to the U.S. AIDS epidemic? *Science , 305,* 1243-1244.

Kennedy, D. (Ed.). (2003). "Viewpoint: Historical Essays." *Science, 298,* 1726-1731. Provides brief articles on AIDS and the discovery of HIV, which were written by Stanley B. Prusiner, Luc Montagnier, and Robert Gallo.

Little, S. J., Holte, S., Routy, J.-P., Daar, E. S., Markowitz, M., et al. (2002): Antiretroviral-drug resistance among patients recently infected with HIV. *New England Journal of Medicine,* 347: 385-394.

Renault, B. (Ed.). (2003). Twenty years of HIV science. [Special issue]. *Nature Medicine, 9.* This special edition of *Nature Medicine* contains a series of articles on HIV/AIDS.

Sepkowitz, K. A. (2001). AIDS—The first 20 years. *New England Journal of Medicine, 344,* 1764-1772.

Steinbrook, R., & Drazen, J. M. (2001). AIDS—Will the next 20 years be different? *New England Journal of Medicine, 344,* 1781-1782.

POPULAR BOOKS

Cohen, J. (2001). *Shots in the Dark: The Wayward Search for an AIDS Vaccine.* New York: W. W. Norton.

D'Adesky, A.C. (2004). *Moving Mountains: The Race to Treat Global AIDS.* New York: Verso Books.

Garrett, L. (1994). *The Coming Plague: Newly Emerging Diseases in a World Out of Balance.* New York: Farrar, Straus & Giroux.

Hill, S. A. (2005). *Emerging Infectious Diseases*. M. A. Palladino, (Ed.). San Francisco: Benjamin Cummings.

Hunter, S. (2003). *Black Death: AIDS in Africa*. New York: Palgrave Macmillan.

Shilts, R. (1987). *And the Band Played on: Politics, People, and the AIDS Epidemic*. New York: St. Martin's Press.

Note. All the Web sites and links presented in this booklet were last accessed and verified for accuracy on October 1, 2004.